CENTRAL AMERICA

AND THE

TRANSIT BETWEEN THE OCEANS.

CENTRAL AMERICA

AND THE

TRANSIT BETWEEN THE OCEANS.

BY

MARMADUKE B. SAMPSON.

REPRINTED FROM THE WESTMINSTER REVIEW,
FOR APRIL, 1850.

NEW YORK:
S. W. BENEDICT, No. 16 SPRUCE STREET.
1850.

CENTRAL AMERICA

AND THE

TRANSIT BETWEEN THE OCEANS.

Numberless signs denote that Central America will henceforth be the theatre of some of the most remarkable changes to be wrought by advancing civilization. Three years back its capabilities and distinctive features were almost wholly unknown to the general public; and such volumes as existed with regard to them, were read with no more active interest than would have been excited by travels in Persia or Dalmatia, or any other country with whom the chances of our establishing an immediate and vital intercourse might be most remote. Now, however, there is no quarter of the world to which attention is more actively directed. Statesmen, merchants, navigators, colonizers, and the students of natural science, are all alike awakened to the importance of its future prospects; and, as a consequence of the demand thus created, books and maps are supplied by our geographical publishers, involving an amount of minute detail, which enables us, we may believe, to form a far more accurate knowledge of each point of the territory, than is possessed by one out of a hundred, even amongst the most intelligent of its natives.

Until now, notwithstanding the almost solemn charm that has invariably been felt in its contemplation, the idea of a communication between the Atlantic and the Pacific, has never been anything more than an interesting engineering problem. That it could be effected without any serious difficulty, has, however, in the face of appearances to the contrary, for many years been placed beyond all real doubt; and the actual point in which projectors have failed, has been simply in convincing the capitalist that it would pay. Appealed to on the strength of sublime estimates of the influences of the enterprise on the destinies of the world, unaccompanied by any data on which reliance could be placed with regard to the per centage in the shape of future receipts, men of business could not be warmed into enthusiasm. In reply, therefore, they have always professed a fear of its impracticability; and, as this was stimulated by the circumstance of each projector abusing the routes proposed by his rivals, it at last became a received belief. They saw all the glory of the project; would be willing to run all necessary risk for its consummation; but the thing was impossible. With a demonstrable dividend before them, every shadow in the shape of a mechanical difficulty would have disappeared.

But the discovery of California has now settled the question of a profitable result; and, in a much shorter time than most persons in England are even yet prepared to expect, not merely a communication, but a choice of communications, is certain to be opened up. These will be respectively at Panama and Nicaragua; the former by railway and steamboat in the first instance, and ultimately by railway entirely; the latter, chiefly by steam-boat in the first instance, and ultimately by a complete canal both for steam-boats and sailing-vessels.

The Panama line is promoted by Howland and Aspinwall, of New York. It is to consist of a railway from Navy Bay on the Atlantic to Panama on the Pacific, at an estimated cost of $5,000,000, or £1,000,000 sterling. At the commencement, however, a portion of the road, consisting of about twenty-two miles on the Pacific side (from Panama to Gorgona), will be constructed and put into operation, and the rest of the transit will be effected by steamers running forty-five miles on the Chagres river, which is navigable at all periods of the year for vessels of light draught. The work, it is estimated, may thus far be completed for £200,000, and the shareholders will be in the receipt of revenue while the remainder is being constructed. The full capital for this portion has been subscribed at New York; the entire line has been surveyed, and the grading of the distance from Panama to Gorgona already contracted for at the price of $400,000, (£80,000), which is within the original estimate. The grant to the Company by the Republic of New Grenada gives them an exclusive privilege for forty-nine years, subject to a right of redemption by the Republic at the end of twenty years, on payment of $5,000,000; at the end of thirty years on payment of $4,000,000; and at the end of forty years on payment of $2,000,000. This privilege is to date from the completion of the road, for which eight years are allowed; and it is accompanied by a concession of exclusive harbour rights at the ports on each side, and also of the necessary land throughout the line, besides 300,000 acres in perpetuity, for the purposes of colonization. The Company are likewise to be allowed to import iron and whatever may be necessary for the construction of the road, free of duty, including all articles of provision and clothing for the workmen. They may also call upon the Government to furnish them the assistance of three companies of sappers; and the only obligation imposed as to the character of the road, is that it shall be capable of transporting passengers and merchandise from one ocean to the other in the space of twelve hours.

The parties by whom the survey of the Panama route was effected, instead of encountering the formidable difficulties that had been anticipated, found that they could lay down a line which would not exceed forty-six miles in length, with a summit of less than 300 feet above the sea, and with curvatures, having nowhere a radius of less than 1,500 feet. Their explorations were extended over the whole of that part of the Isthmus, so as to insure the one true point, and there can be no doubt that this has now been selected. Another difficulty which has always been represented as no less formidable than the natural impediments of the route, namely, the procuring a proper supply of laborers able to stand the climate, has also been proved to be delusive. The parties who have contracted for the grading of the twenty-two miles on the Pacific side are, it seems, two American engineers, who have been employed for the last five years in the State of New Grenada, in forming a canal ninety miles long, to connect two branches of the Magdalena river, and which they have completed entirely with native labour. They can bring with them a large number of these workmen, whose training, although at first difficult, was ultimately quite successful; and there is reason also to believe that arrangements for foreign labour might be made, since the experience of the corps employed in the survey of the railway, consisting of forty engineers and assistants, was not discouraging with respect to climate.

The explorations for this survey have led to the discovery of large groves of mahogany, and rich mineral deposits, "the knowledge of which," it is represented, "will be highly important to the company in locating lands under their grant;" and with regard to the proposed terminus of the railway on the Atlantic side, on the island of Manzanilla, in Navy Bay, we have the following characteristic speculations, which might, perhaps, be taken as nothing more than a rhapsody, were it not for our experience of the way in which these American visions are apt to produce their own realization.

"The harbour is accessible at all seasons, and with any wind perfectly secure, and capable of containing 300 sail. Of the island, Mr. Norris, the chief engineer of the Chagres division, says, 'in ten years I predict the whole will be covered with houses, and the inhabitants enjoying perfect health, with every luxury of a southern clime.' He adds, 'I do consider it the most eligible and perfect site for a city of any size I have ever seen.'"

The second line, which may now be considered definitively arranged, is that of a ship canal in connexion with the lakes of Nicaragua. This work promises an early commencement, and also a rapid progress. On the 27th August last a contract was made between the State of Nicaragua and the Atlantic and Pacific Ship Canal Company, of New York, by which all the exclusive privileges necessary for the undertaking were conferred. According to the contract, the canal is to be completed within twelve years, unless prevented by fortuitous occurrences; and, upon failure of this stipulation, such part as may have been constructed is to be forfeited to the State. The Company agree to pay the State $10,000 for the ratification of the contract; $10,000 more annually till the completion of the work; and to make a donation of their stock to the amount of $200,000. When finished, the State is to receive one-fifth of the net profits for twenty years, and afterwards one quarter. There is also a stipulation that it is to have 10 per cent. on the profits of any minor line of communication between the two oceans which the Company may open up during the time they are engaged on the grand canal. On the other hand, the privileges bestowed are, not only the exclusive rights for constructing the canal, but also the exclusive right of inland steam navigation; grants likewise are to be made of eight sections of land on the banks of the canal, each section to be six miles square. The concession, as originally proposed, was for eighty-five years, but it has since been extended in perpetuity.

The first payment of $10,000 has already been made by the company to the Government of Nicaragua; and the general arrangement having been favourably viewed by the cabinet at Washington, there can be no question that the privileges which have been acquired under it may now be regarded as incapable of being upset. It is true that Mr. Barclay, the British consul at New York, has given notice to the Company that in extending the grant to the exclusive right of navigating the river San Juan, the State of Nicaragua has entered into an agreement in regard to places where it has no competence, since "the boundary line of the Mosquito kingdom touches the St. John's river, at the Machuca rapid, about thirty miles below the Lake Nicaragua, from whence to the mouth of the St. John's the navigation belongs to Mosquito;" but this, supposing the English view to be admitted, can in no way affect the main privileges they have obtained. It would necessitate a negotiation on their behalf to obtain from the King of Mosquito, or, in other words, from the English Government, a guarantee of those rights on the San Juan which it is not in the power of Nicaragua to grant; and in this it is to be inferred there would be no obstacle, since it would be impossible to refuse the application, so long as the Company are ready to bind themselves that the route, when constructed, shall be open, on fair and equal terms, to the whole world, and that the power of holding its stock and of participating in its management shall likewise be free to all parties. Indeed, it is believed that the question has already been met, both by Lord Palmerston and Mr. Abbott Lawrence, in a spirit which will speedily lead to a joint guarantee on the part of England and the United States of the neutrality of the whole line. As to political difficulties, therefore, so far as the promoters of the canal are concerned, there are actually none. A short, although a vexatious delay—for even a few months' impediment to such an undertaking would be an evil full of reproach—is all that could arise out of the uncertainty at present existing on these points. Neither England nor the United States would like it to form a part of their history, that the human race had been kept, for two or three or more years, from witnessing the junction of the Atlantic and the Pacific, because their foreign ministers had been unable to adjust a wrangle as to whom the right belonged of conceding some seventy miles of the now desolate track through which the work would pass.

In the projects for the Nicaragua Canal hitherto put forward, the estimated cost has been £4,000,000, the actual outlay being reckoned at £3,600,000, and the remaining £400,000 being allowed for casual expenses. These calculations were professedly made on the high scale of the Caledonian Canal, where the expenditure was notoriously reckless, and at a period when the mechanical facilities for such undertakings were very imperfect as compared with the present time. They were based, however, upon the surveys of Mr. Baily, which are regarded as having been made with a degree of conscientious care entitling them to the highest credit. Hence, in the calculations in question, there are no existing points of engineering difficulty which were not comprised; and it may accordingly be inferred, that if they were under the mark, the deficiency was simply caused by not allowing enough for labor, materials, and interest of money. The improvements in mechanical science, and the diminution in the cost of materials during the last ten or fifteen years would, it may at least be assumed, make up as large an amount on the other side; and allowing, therefore, for the invariable experience which attends all estimates, there is now no reason to suppose that, under any circumstances, the cost could exceed the total that has been contemplated. This is somewhat less than half the amount that has been expended for the Dover Railway, and about two-thirds of the expenditure for the Brighton.

The revenue, which was calculated years ago when the original schemes were propounded, was taken upon 900,000 tons, and the contemplated toll per ton was 10s. for European, and 20s. for United States vessels; the whole producing about £600,000 a year, which, after leaving two per cent. for maintenance, and one per cent. for sinking fund, would yield a return of twelve per cent. on the capital.

An examination of these estimates, however, produces no conviction of their correctness. All the materials on which they are founded are extremely vague; too much reliance was placed on the change of route to India, and the proposed difference in toll to American vessels would, moreover, never be tolerated. This difference was suggested on the idea, that as the average saving of time to United States vessels would be two months, and to European vessels only one month, toll should be exacted in a proportionate ratio; but it would introduce an entirely new principle into the universal system of navigation dues, and public charges of all kinds, and one that would be found as impracticable as it would be unjust and absurd.

But since these statements were made in 1835, the traffic with South America has greatly increased, and Australia and New Zealand have been growing in importance. Still, even with these changes, capitalists would possibly have regarded the experiment with hesitation. At all events, it would have been one of anxiety. It is, as we have already observed, the discovery of the gold mines in California that has alone altered the whole aspect of the affair, and rendered it such as will be pursued with eagerness.

While the Panama Railway will take the whole of the passengers for the western ports of South America, the Nicaragua route must command the entire traffic to California the moment it shall be rendered practicable, even by a mixture of water and land conveyance. On the completion of the canal, it will of course, in addition, monopolize all the shipping trade between the two oceans, but some of its most startling results will be witnessed long before that period. The distance saved by the Nicaragua route in the journey to California, as compared with the Panama, is sufficient to prevent the possibility of competition; and apart from this, the attractive features of the former are such as to give it an unquestionable superiority. Now, the emigration to California from the United States has recently been 6,000 or 7,000 persons each month, or at the rate of 80,000 per annum; and one peculiar feature of that emigration seems to consist in the fact that, at whatever rate it may continue, it will always be of a shifting kind,—that is to say, there will always be a tide of persons both going and returning. Gold digging can only be carried on for about five months out of the twelve, and during the idle season it would be far more economical to return to the States than to live at San Francisco. The operation itself is also one which men are only disposed to pursue temporarily, so that after a little while each miner is content to return and to leave his place to a new comer. This has been particularly exemplified during the past autumn,—both the influx and efflux of passengers having been enormous; and that the efflux was not caused by persons who were returning in disappointment has been abundantly shown, by the fact of their reäppearance in the United States not having led to the slightest diminution in the number of those who were still eager to emigrate. According to the last accounts, ships of a still larger and finer class than those hitherto employed were being placed upon the service, and every ticket in the three lines of steamers had been taken up to May or June. There is consequently ground to calculate on a constant stream both ways. The certainty of this is indeed demonstrable. At present the average to each miner is at least five dollars per day; and supposing the supply of gold to continue at this rate, population must steadily flow in until the rate of wages for a similar day's labor, after making allowance for the expense of passage money, &c., shall have been equalized throughout the world. Each mail repeats the story that no one in the country doubts the supply to be comparatively inexhaustible. By the last advices, Colonel Fremont, moreover had discovered, between San Francisco and Monterey, a vein in the mountains which yielded the extraordinary proportion of one ounce of gold to twenty-three pounds of rock. Quicksilver and silver mines were also waiting only for machinery and labor. Instead of a diminution of activity in this direction, everything therefore indicates an increase.

However much we may be disposed to distrust the twelve per cent. estimate of those who in former years proposed the execution of the work, we must under these circumstances admit that there can be little fear of its present results. We must look not only at the traffic which is even now before us, but we must take into account its natural increase from the greater cheapness and rapidity of the new route. We must also look at the growing importance of Oregon, and to the certainty of the crowd of small steamers that will rapidly accumulate on the Pacific, from the smoothness of its waters and the abundance of the easily worked coal of Vancouver's Island.

At the same time, although the view is thus bright, there is no great likelihood that it will attract any amount of English money. Faith, the great element of all enterprises, has been destroyed in this country for many years to come ; and not only is there no disposition to enter upon the scheme among ourselves, but there is a strong tendency to suppose that others would be equally timid, and to doubt if the Americans would or even could carry it out without "the aid of British capital." Such has been our step from the sublime to the ridiculous, that we have come to look upon the expenditure during the next twelve years of a sum of 4,000,000*l.* (which is a little more than half the amount of the railway calls for the month of January, 1847), upon the grandest public work that mankind has ever contemplated, as something that is really appalling from its temerity, and that is only to be carried out by a congress of capitalists from all the nations of the earth. In the United States, however, the feeling is very different; and every year vast works are quietly undertaken there, and carried to completion in a way which would surprise those numberless people who are too apt complacently to believe that all the world stands still except when funds are sent from London. They have enjoyed prosperity since

1839; and although, of course, after so long a period, their turn for a run of madness must be approaching, there are at present no signs of it, and no apprehensions of its arrival for two or three years. They are quite prepared, therefore, to look confidently at any rational project, however broad, and nothing could be presented to them which would more enlist their commercial aptitudes, their hard energy, and practical benevolence, or their patriotic pride. "I would not speak of it," said one of their writers, a few years back, "with sectional, or even national feeling; but if Europe is indifferent, it would be glory surpassing the conquest of kingdoms to make this greatest enterprise ever attempted by human force, entirely our own."

We may rely, therefore, that the day is gone by when the undertaking could be neglected for want of funds. If carried out entirely by capitalists in the United States, it will probably be pushed forward with less rapidity than would otherwise be the case; but this will be far more than compensated by the exercise of greater economy and certainty. Meanwhile, steps have already been taken for ascertaining what will be necessary to render the route immediately available for passengers, and for placing steamers upon the river San Juan, and the lakes. The Chairman of the Company—a Mr. Vanderbilt, who it is said has been more largely and profitably connected with steam navigation than any other citizen of New York—started some months back on a personal survey of the entire district; and, as he and his friends are understood to be prepared to subscribe for a very considerable proportion of the required capital, a report may be expected, in which, contrary to English usages, the interests of the stockholder will be consulted before those of the engineer.

The precise course which will be taken by this canal, whenever it may be completed, is still in some parts uncertain; not from any question of great difficulty, but from the fact of three modes presenting themselves for the exit from the lake to the Pacific, from which a selection is to be made. From Greytown (or San Juan) on the Atlantic, the course for 104 miles is by the river San Juan; the Lake of Nicaragua is then entered, and it is the best route from this lake to the Pacific that remains to be determined. The line contemplated and surveyed by Mr. Baily, was from the south-western point of the lake to the port of San Juan del Sur, the extent of which would be fifteen miles, with an elevation to be overcome, in one part, of 487 feet. Another route which has been proposed but not surveyed, is from the same part of the lake to the port of Las Salinas, lying within the boundary claimed by Costa Rica, which would be about the same length, but which would not, it is said, present a greater elevation than 270 feet; and a third proposal is, to proceed from the northern part of the lake by the river Tipitipa, twenty miles in length, to the smaller lake called Lake Leon, and thence by a canal of eleven miles through a district which is alleged to offer no greater rise than fifty-one feet, to the river Tosta, which communicates at eighteen miles distance with the well known port of Realejo. At present, opinion seems to tend toward the last named course, as the one that would be most advantageous; but it would be idle with the limited materials now before us to speculate upon the point, since we shall soon be furnished with detailed statements prepared by practical men, who have entered upon the task of selection with all their interests enlisted in the matter, and with a thorough perception of the way in which all views regarding it must henceforth be adapted to meet most favourably the altered circumstances of commerce that have arisen in connection with California. The port either of San Juan del Sur, or of Las Salinas would seem to be in some measure the most desirable, if the trade with South America, Australia, and New Zealand, were made the predominant consideration; but as respects Mexico, San Francisco, Oregon, Vancouver's Island, the Sandwich Islands, and the Indian Seas, which will be undoubtedly by far the most extensive region of traffic, Realejo is the best. Indeed, it is possible that with this view a yet more northerly terminus may be selected, and that in preference to that port, the line may be made to run into the Gulf of Fonseca.

The certainty of these two routes of Panama and Nicaragua being speedily carried out in a more or less perfect degree, places the rapid settlement of Central America beyond all doubt; and hence gives to all personal descriptions of the country, such as those which have been furnished by Mr. Baily and Mr. Byam, an interest that comes home to our daily business. Let the reader imagine what must be the effect even of an annual transit of 50,000 or 100,000 adventurous and well-informed people through a strip of country scarcely one hundred and fifty miles broad, yet commanding the ocean intercourse with Europe on one side and with Asia on the other, favorable to health, and abounding, at the same time, owing to the inequalities of its surface, with every natural product that can be found distributed elsewhere, between Scotland and the tropics, and an impressive idea of its coming destiny will be awakened; but let the glance be carried further, to the period of the completion of the canal, and then let it be remembered that within this strip of land lie two calm, yet deep and extensive lakes, that seem, as we look upon them in the map, like huge natural docks in the centre of the world, intended to receive the riches of a universal commerce; and in the contemplation of what is yet to be realized, the mind will almost beat with impatience against the slight barrier of time which yet remains between us and its accomplishment.

That Central America possesses inherently all the essentials to attract a dense and vigorous population, is a fact that has rarely been doubted by those Europeans or Americans who have visited the country, and all the publications before us tend to confirm it. The researches of Mr. Stephens showed that it had been largely peopled by an aboriginal race of a remarkable character, and the size of its towns and its architectural remains give evidence of comparative prosperity under the old Spanish dominion. Leon, the principal city of Nicaragua, was formerly noted for its opulence, and once contained 50,000 inhabitants, who were among the most peaceful and industrious people in the country; while it has now, it is said by Mr. Baily, not more than one-third of that number, and half the place is in ruins. This is simply owing to the wretched revolutionary contests that have gone on without intermission since the declaration of independence, and which are invariably got up by a handful of military vagabonds, who would be swept away in the course of four-and-twenty hours, or who, rather, would never dare to show their faces if a hundred Englishmen or Americans were in the district to stimulate the well-disposed to confidence.

"The fact is," says Mr. Byam, "that every revolution effected in all the republics, from Chili to Mexico, is brought about by such a mere fraction of the population, that it seems a wonder to an Englishman that the great majority do not arise and speak out—'We wish to be quiet; we do not want revolution and murders; nor do we wish to be subjected to forced contributions of money, cattle, and personal service; and, above all, we are nine out of ten in number against your one; and the great majority will not consent to be plundered by the small minority, who are only dissolute ruffians.' "

If the reign of peace were established, (and even now it may be considered that such is almost the case, for after the present year we shall hear no more of disturbances in Nicaragua,) the progress of the country, apart from the effects of a large European immigration, would of itself be steady and considerable.

With regard to health, the varied productions of Central America give the best evidence that whenever the country shall be opened up by roads and steam-boats, and all the locomotive appliances of modern science, there will be no condition of person, who may not, by ordinary attention to the natural laws, enjoy in this territory all the physical power of which his constitution may be capable. Wherever it is possible to reach by a few hours' journey, districts in which wheat, barley, and all the ordinary fruits and vegetables of Europe may be grown in perfection, there can be little fear that anything will be wanting in the way of climate to insure the preservation of bodily vigour. Even in its present state, Central America, on the whole, has no bad reputation regarding health, although the advantages offered by its configuration in enabling the inhabitants to vary their climate according to their requirements, might as well not exist, since roads can scarcely be said to be known, the best rate of progress being about twenty miles a day, and mule paths through thick woods, without resting-places at night, being usually the only features of a traveller's track. Yet, on the banks of the San Juan, and in other parts of Nicaragua, there are elevations that would afford the most beneficial sites for farms and residences; while in Costa Rica, San Salvador, and indeed in all the states, table lands more or less abound, where any condition of climate may be obtained in a few hours. In Guatemala may be seen fields of wheat and peach-trees, and large districts "resembling the finest part of England on a magnificent scale." Valuable mineral and thermal springs are likewise distributed over the various localities, and there are other adjuncts of a curative kind, which may possibly be found to yield extensive results, and to present even a temptation to some classes of invalids. Amongst these is an animal called the manatee, between a quadruped and a fish, about ten feet long, weighing from 500 to 800lbs., affording excellent food, and possessing a medicinal quality, apparently analogous to the cod-liver oil, it being alleged to be strikingly effectual as a speedy cure for scorbutic or scrofulous disorders. "The blood is said to become purified, and the virulence of the complaint thrown to the surface of the body, quickly disappears."

"Although Central America," observes Mr. Baily, "occupies the middle space between the equator and the tropic of Cancer, consequently lying within the torrid zone, the temperature may be said to be relatively mild, and, taken altogether, it undoubtedly is salubrious;" and this it must be remembered is the testimony of an English officer, who has resided in the country from choice during the best part of his life. The places most prejudicial to health lie on the northern coast and the Mosquito shore, where endemic and intermittent fevers are not unfrequent. The Pacific coast is exposed to a temperature equally high or nearly so; but is much more salubrious, and seldom visited by epidemic or contagious diseases.

In point of natural riches, Nicaragua and Costa Rica have usually been spoken of amongst the various States as possessing the most abundant resources, but they all teem with rewards for industry, such as is almost unknown in any other part of the globe; and upon a review of the claims of each state in this respect, it is hard to decide which has the greatest capabilities. In the plain of Nicaragua the fields are covered with high grass, studded with noble trees and herds of cattle. Cocoa, indigo, rice, Indian corn, bananas, and cotton are here produced, and mahogany, cedar, and pine abound in the forests. On the eastern side of the lake there are cattle farms on which are herds of from 10,000 to 40,000 oxen, bulls, and cows. Horses and mules are bred for riding and for burden. Sheep are reared on the upper plains, and swine are kept for flesh. A planter from one of the West India islands stated as his conviction, in reference to the district round Lake Leon, that, provided he could get the same amount of labour, he could manufacture sugar at one-fourth its cost in the West Indies. At present it is sold in Nicaragua for three half-pence per pound. Leaving the lakes, and descending the San Juan, each bank of the river is covered with valuable wood, of all sizes and descriptions, and the land is of prodigious fertility. With regard to the mining wealth of Nicaragua, Mr. Byam made some interesting observations, but the miserable state of the laws, and the spoliations of the government, prevented him from carrying on the enterprises in connexion with it, to which he might otherwise have been tempted. The copper ores he met with were almost all uncombined with sulphur or any other substance that requires calcining to be got rid of, and they were consequently such as might be smelted in a common blast furnace, with the aid of equal quantities of iron-stone, which lies in abundance on the surface of all the hilly country. He found also silver mines, consisting of fine broad, but rather irregular veins, the ore of which was combined with a great quantity of sulphur and a large proportion of lead. For the want of a silver assaying apparatus he could not get a good assay; but with the means in his power he could produce about fifteen marcs of silver the ton. "The mineral riches that are deposited in the bosom of these mountains," he adds, "are no doubt very great; but the working of the mines is so difficult, from the ignorance of the workmen who have to be taught everything, their invincible idleness, and the vacillation of the government, that I believe it will be long before anybody will be found to advance capital for prosecuting such a forlorn undertaking." This, however, was written when there seemed no gleam of hope for the resuscitation of the country.

Among the numerous products which Mr. Baily points out as offering temptations to the cultivator, are fruits of various kinds, indigo, and the mulberry for silk-worms. Fruits of the country, it is said, are sufficiently plentiful, as well as oranges and lemons, which are excellent. Vegetables and garden produce are scarce about Leon, but they might be raised in all parts in great perfection; but not being considered of so much importance by the natives as by foreigners, they are unattended to. With regard to indigo, the quality already produced will bear "an advantageous comparison with the finest of any country whatever;" and no part of Central America is better suited to a more extended cultivation of it; yet, with all the advantages that are presented, few efforts are made to increase the annual growth. "The cause of this neglect is mainly attributed, in recent years at least, to a diminution of capital, and possibly, in no small degree, to an apathetic indifference to the future, consequent upon the misfortunes arising from a continued series of internal discords that unhappily have paralyzed all industrial pursuits." Of the *morus multicaulis* it is remarked, "the mulberry for silk-worms grows remarkably well, and the climate appears to be congenial to it in all respects. Hitherto, little or rather no advantage has attended the cultivation, chiefly from want of attention and requisite experience. Were these deficiences supplied, and the business carried on with energy and skill, a large quantity of silk could be produced. How profitable such an article of commerce would be to proprietors needs no demonstration."

The impossibility of any profitable cultivation either of these or of any other articles, except for home use, in the present state of the country, will easily be understood from the circumstance that the rate of conveyance for merchandise and produce seems to be about two or three dollars per cwt. for every hundred miles; while the possibility of transporting it even at this charge, and at a speed of about twelve miles a-day, depends upon the supply of mules that may be available. It is likewise to be observed, that agricultural implements are almost wholly wanting. The plough, the harrow, the scythe, the sickle are not found on the farm; and the hoe, and the machete are the only substitutes for them. Under these circumstances, the rearing of cattle is almost the only branch of occupation that is carried on to any extent; but from the difficulty of transit to the markets, where they would be in demand, a good bullock is only worth from four to six dollars, and abundant pasturage yet remains unappropriated. "From a fertility of soil capable of maintaining millions, little more is now drawn than the sustenance of 250,000 inhabitants; but," Mr. Baily truly observes, "when, by increase of population, a greater command of capital, more intelligence of agriculture, commerce, and political economy, which in process of time will creep in, the beneficence of nature shall be looked upon as incitement to industry; and encouragement shall be given to raise produce of exportable value and general demand, Nicaragua will be converted into a region of immense wealth."

Of the other four republics of Central America by which Nicaragua is surrounded, namely, Costa Rica, San Salvador, Guatemala, and Honduras, only a few more words are necessary. In Costa Rica, as in Nicaragua, the soil is singularly productive; and all the articles peculiar to inter-tropical regions are grown in abundance, excepting cochineal, cotton, and the vine, which are liable to be destroyed by the periodical rains. Coffee is the staple export, and as well as indigo, tobacco, and cocoa, which are also produced, is remarkable for its quality. Woods, drugs, grain, fruits, poultry, and a variety of miscellaneous articles likewise form part of the commerce of this little republic. Some gold mines exist, and are at present being worked, although without any very extraordinary results. Copper and coal are likewise found, but these of course have been neglected. The population amounts to 100,000 inhabitants, of whom only 10,000 are Indians. The trade is now almost exclusively carried on with England in British bottoms; but the shipments taking place on the Pacific side, the tedious route by Cape Horn is a serious drawback. In 1848, the exports consisted of 150,000 cwt. of coffee, estimated at $6 on board; of about 10,000 ox and cow hides; of a considerable quantity of mother-of-pearl, Nicaragua-wood and sarsaparilla, and of a small number of pearls; the total estimated value being $1,000,000. San Jose, the capital, is 4,500 feet above the level of the sea, and from this a cart-road of seventy-two miles forms the communication with the port of Punta Arenas on the Pacific. The great want of this republic has been a communication with the Atlantic, so as to save the long navigation by Cape Horn, and the government are now proceeding vigorously with a road of 66 miles from San Jose to the Sarapiqui river, which runs into the San Juan, and will thus furnish the opening that is desired. Costa Rica is the only one of the republics of Central America that for any lengthened period has been free from anarchy, and the result is that she is steadily advancing to prosperity, and that a treaty of amity, commerce, and navigation was concluded with her by Great Britain on the 20th February last. She has at present a minister in London, Senor Molina, who is understood to be the writer of a very intelligent pamphlet on her resources, which has lately been published. By some notices in the French paper, *La Presse,* we also remark that a considerable grant of land has been made to a gentleman in Paris, for the promotion of colonization in a part of the state situated in the Gulf of Dulce, on the Pacific.

The state of Salvador is the smallest of the five republics, but relatively the most populous, the number of her inhabitants being 280,000, and her natural resources and position on the Pacific being calculated to admit of the utmost prosperity. She has, however, been incessantly ravaged by civil discord; and it is only about two months since a large body of her people joined some insurgents in the neighbouring state of Guatemala, with the view of overturning the government in that country; while we have also seen that it has just been necessary for an English ship of war to blockade her ports in order to exact restitution for a fraudulent seizure of the property of British subjects. The chief production of San Salvador has been indigo; but she has the highest capabilities also for tobacco, cotton, sugar, and coffee. The mineral workings have been considerable. Gold was formerly, and still is extracted; and rich silver mines, which were once wrought, are known still to be valuable. "But for many years past, no one has wished to be thought rich enough to work a mine, lest he should be called upon to pay exorbitant contributions to the exigencies of the state." Copper and lead exist in different parts; and, near a town called Matapam, a very superior iron ore is abundantly obtained, which, looking at the price commanded by all foreign iron, might, it is believed, be made to yield very profitable results. On that part of the coast of Salvador, extending from Acajutla to Libertad, is collected the article known in commerce as the balsam of Peru—a name it erroneously received from having been first shipped to Callao, and thence transmitted to Europe.

The state of Honduras has an estimated population of 236,000, and, although possessing excellent capacities both in soil and climate, is chiefly remarkable as a mining district. It contains gold and silver mines, long neglected, owing to the ruin and insecurity occasioned by constant revolutions. Lead and copper, also, in various combinations, as well as opals, emeralds, asbestos, and cinnabar. An abundance of timber and dye-woods is likewise presented, and vast herds of almost profitless cattle range over lands that are otherwise unoccupied.

Guatemala has a population of 600,000, and nearly all the surface of the state is mountainous. In point of salubrity, extent of available lands, and quality of the soil and climate, the finest field for European immigration is perhaps to be found in this quarter.

"Maize and wheat," it is said, "are abundant, and of superior quality; rice is excellent; the tropical fruits and vegetables are good, and in great variety; and the produce of leguminous plants is equal to the best of that grown elsewhere. All European fruits and garden-stuff grow kindly; and if the Indians, who are the only cultivators, were better instructed in the art of horticulture, they would be carried to an enviable degree of perfection; in fact, but few regions are so well endowed with the capabilities of producing all that ministers to the comforts as well as luxuries of life. Of things more important in a commercial view, cochineal at present holds the first rank; to which may be added cacao, tobacco, sugar, coffee, silk, cotton, wool, and a numerous list of minor articles."

In glancing at these leading characteristics of the various states of Central America, the reader will speedily have arrived at the conclusion that, in the hands of Anglo-Saxon settlers, they would long ere this have ranked amongst the most beautiful and prosperous portions of the earth. But until now there has been work for the race in higher latitudes, and it will be from the present year that their rise will date. The nature and rapidity of that rise will, we believe, be such as has never yet been witnessed in any analogous case. Emigration from the United Kingdom has hitherto been confined to swarms of the poor, going out to fight the battle of life in untilled solitudes, where they might best enter upon it with unburthened limbs; and although their progress has been wonderful, and they have caused cities and states to rise up as if by magic, there have still been rough elements in the whole proceeding which have left room for us to contemplate the possibility, under more favorable circumstances, of an equally rapid progress, coupled with a far higher and finer civilization. All separation of classes is bad, and the true system of emigration, where the temptations for it exist, is that, where the rich and the poor, the educated and the uneducated, go together. But the rich and intelligent will go only from choice, and they demand as inducements a brighter sky, a more genial climate, and facilities of communication.

New Zealand, from its possession of the two first recommendations, has already attracted many, but its distance and solitariness are fatal objections. Central America promises to fulfil every required condition. In a short time the active spirits from New York and Boston, who are even now infusing new life and hope into Jamaica, from merely calling at that island in their way, and stirring up its inhabitants to the resources at their feet, over which they have hitherto blindly moped, will have displaced the spirit of anarchy by that of enterprise. There will then be abundant work for the laborer, and temptations for all classes even to the highest. The merchant can seek no broader field than one where he can deal with the meeting commerce of two worlds, together with every variety of teeming produce at his own door. The agriculturist, the fisherman, the miner and the engineer will likewise find greater stimulants and rewards than can be met elsewhere. The artist will be incited by scenery which in its condensed grandeur and prolific beauty from the mountain Ysalco in Salvador, which burns incessantly as a natural lighthouse on the Pacific, to the frosty table-lands of Guatemala, combines, like the soil and the climate of the country, every feature that is otherwise only to be witnessed by extended wanderings. The naturalist, the geologist, the astronomer, and the antiquarian will here also have a new range; and the man of so-called leisure, who in his way unites the pursuits of all, will proportionably find the means of universal gratification.

And in the narrow confines which hold these advantages the people of every land and government are destined to meet on common terms. The Russian from Behring's Straits, the Chinaman, the African from Jamaica, the New Zealand sailor, the Dutchman from Java, and the Malay from

Singapore will mingle with the Mestizoes and Indians of the country, and each contribute some peculiar influence which will be controlled and tempered to the exaltation of the whole by the predominant qualities of the American, the Englishman, and the Spaniard. Is it too much to suppose, that under these circumstances, a people may arise whose influence upon human progress will be of a more harmonious and consequently of a more powerful kind than has yet been told of?—that starting at the birth of free-trade, and being themselves indebted to a universal commerce for their existence, they will constitute the first community amongst whom restrictions will be altogether unknown; that guaranteed in their independence by Great Britain and the United States, and deriving their political aspirations from a race amongst whom self-government is an instinct, they will practically carry out the peace doctrines to which older nations are only as yet wistfully approaching; that aided and strengthened by the confiding presence of people of every creed, the spirit of Christian toleration will shine over all, and win all by the practical manifestation of its real nature; and finally, that the union of freedom, wisdom and toleration may find its happiest results in the code of internal laws they may adopt, so that amongst them, on the luxuriant land hitherto made desolate by the sole principle of bloody retaliation, the revengeful taking of human life may never be known; and that they may be the first to solve the problem —if amongst those who profess Christ's doctrines it can be called a problem—of coupling the good and reformation of the offender with the improvement and safety of society, and the exercise towards both, not of a sentimental, but of a philosophical and all-pervading love.

www.ingramcontent.com/pod-product-compliance
Lightning Source LLC
LaVergne TN
LVHW021200120826
845150LV00011B/2293

* 9 7 8 1 4 1 8 1 9 4 8 9 5 *